YOUR KNOWLEDGE HAS VALUE

AF300935

- We will publish your bachelor's and
  master's thesis, essays and papers

- Your own eBook and book -
  sold worldwide in all relevant shops

- Earn money with each sale

Upload your text at www.GRIN.com
and publish for free

GRIN

The Importance of Litter in Agroforestry. Definitions, Dangers and Recommendations

Inemesit Eniang

Bibliographic information published by the German National Library:

The German National Library lists this publication in the National Bibliography; detailed bibliographic data are available on the Internet at http://dnb.dnb.de.

ISBN: 9783346911148
This book is also available as an ebook.

**MICHAEL OKPARA UNIVERSITY OF AGRICULTURE, UMUDIKE
COLLEGE OF NATURAL RESOURCES AND ENVIRONMENTAL MANAGEMENT
DEPARTMET OF FORESTRY AND ENVIRONMENTAL MANAGEMENT**

BEING AN ASSIGNMENT WRITTEN FOR THE PARTIAL FULFILMENT OF COURSE FFA 211

(AGRO FORESTRY SYSTEM)

TOPIC:

IMPORTANCE OF FOREST LITTER IN AN AGROFORESTRY

ECOSYSTEM

WRITTEN BY:

ENIANG, INEMESIT EDEM

THE DEPARTMENT OF FORESTRY AND ENVIRONMENTAL MANAGEMENT

MAY, 2023

1

Abstract

Agroforestry is a sustainable land management system that combines the cultivation of crops with the growth of trees and shrubs. In such systems, the role of leaf litter, the organic matter that accumulates on the forest floor via leaf falls, is of utmost importance. Litter plays a crucial role in enhancing soil fertility, improving water infiltration and retention, promoting nutrient cycling, and fostering better plants performance as wee as rich biodiversity. This paper explores the significance of litter in agroforestry ecosystems, highlighting its multifaceted benefits and underscoring its role in promoting sustainable agricultural practices.

Introduction

Agroforestry is defined as a land-use management system that combines the cultivation of trees or woody plants with agricultural crops and/or livestock on the same piece of land. It is a sustainable agricultural practice that aims to optimize the benefits of both agriculture and forestry by integrating them into a unified system (Kitalyi et al.2013).

They went on to state that in agroforestry systems, trees are intentionally planted alongside agricultural crops or pastures in a designed and managed manner. The specific configuration and arrangement of trees and crops can vary depending on the goals of the system and the ecological conditions of the area.

Thus, the concept of agroforestry recognizes the numerous benefits that trees can provide to agricultural production and the environment. Some key benefits of agroforestry include:

Increased biodiversity: The presence of trees in agroforestry systems enhances habitat diversity, supporting a variety of plants, insects, birds, and other animals. This increased biodiversity can contribute to improved ecological resilience and better pest control.

Soil conservation and fertility: Trees help prevent soil erosion by reducing erosive power of wind and water runoff. They also contribute organic matter to the soil through leaf litter and root decay, improving soil fertility and structure (Ajayi, O. C., et.al 2009).

Nutrient cycling: Agroforestry systems promote nutrient cycling by utilizing trees to capture nutrients from deeper soil layers and cycling them back to the surface through leaf fall and decomposition. This helps maintain nutrient availability for crops and reduces the need for external inputs eg inorganic fertilizer (.Minang, P. A,et al. 2006)

Climate change mitigation: Trees in agroforestry systems sequester carbon dioxide from the atmosphere through photosynthesis, thus acting as a natural carbon sink. They can play a significant role in mitigating climate change by reducing greenhouse gas emissions and enhancing carbon storage (Akimoto, k,et al.2014)

Microclimate regulation: Trees provide shade, windbreaks, and evapotranspiration, which can help moderate microclimatic conditions in agricultural areas. This regulation can reduce temperature extremes, enhance water availability, and protect crops from adverse weather conditions (Nair, P. K. R.,et al.2004).

Diversification and resilience: Agroforestry systems provide a diversified production system, reducing the vulnerability of farmers to market fluctuations and extreme weather events. The integration of trees and crops offers multiple sources of income and ensures a more resilient and sustainable farming system. Agroforestry practices can take various forms, including alley cropping (planting rows of trees with crops in between), silvopastoral systems (combining trees with livestock grazing), and forest farming (cultivating crops under the shade of a forest canopy). The specific choice of agroforestry system depends on factors such as climate, soil conditions, local agricultural practices, and desired outcomes (Minang et al .2006)

Overall, agroforestry promotes the integration of trees into agricultural landscapes, emphasizing the potential for synergistic relationships between agriculture and forestry. By harnessing the ecological benefits of trees, agroforestry offers a sustainable approach to food production, environmental conservation, and rural development. (Montagnini, F., & Nair, P. K. R. (2004)

DEFINITION OF LITTER

Litter refers to the layer of organic material that accumulates on the surface of the soil in natural ecosystems. It is composed of various types of dead plant material, such as leaves, twigs, bark, and plant debris, along with other organic matter like animal excrement and decaying animal remains. Litter plays a crucial role in nutrient cycling, soil health, and overall ecosystem functioning (Wakefield, R. 2016). (See plate 1 and 2)

[Editor's note: This image had to be deleted due to copyright issues.]

Plate 1: A layer of leaf litter in a rainforest site in Akwa ibom State

The components of litter can be broadly classified into two categories: living and non-living.

Living Components:

A. Leaves: Leaves are a major component of litter. They contain nutrients and organic compounds that are gradually released as they decompose, enriching the soil with essential elements.

B. Twigs and Bark: Twigs and small branches that fall from trees, as well as pieces of bark, contribute to the woody component of litter. They take longer to decompose compared to leaves.

C. Flowers and Fruits: Flowers, fruits, and their associated parts that fall to the ground contribute to the litter layer. They contain seeds and nutrients that can aid in the regeneration of plants.

Non-Living Components:

A. Dead Plant Stems: Dead stems and branches from plants make up a significant part of litter. These plant parts gradually break down over time.

B. Plant Debris: This includes broken plant fragments, such as shredded leaves, pieces of broken stems, and fallen petals, which contribute to the overall organic matter in the litter layer.

C. Animal Excrement: Feces and urine from animals, including insects, mammals, and birds, can become part of the litter. They add organic matter and nutrients to the soil.

D. Decaying Animal Remains: Dead animals, including insects, small mammals, and birds, can become part of the litter layer. As they decompose, they release nutrients back into the ecosystem.

The decomposition of litter is facilitated by various organisms, including bacteria, fungi, insects, and earthworms. These decomposers break down the organic matter, releasing nutrients that can be taken up by plants and utilized in the ecosystem. The rate of decomposition depends on factors such as temperature, moisture, litter quality, and the activity of decomposer organisms.

Litter serves important functions in ecosystems, including providing a protective layer for the soil, promoting moisture retention, moderating temperature, preventing erosion, and acting as a source of nutrients for plants. It also contributes to the formation of humus, which improves soil structure and fertility (Blair, J. M., & Parmelee, R. W. 1990).

[Editor's note: This image had to be deleted due to copyright issues.]

(Plate 2)

Litter and Soil Fertility

a. Litter as a nutrient source for plants b. Role of litter in improving soil structure c. Litter's contribution to soil organic matter content

A. Litter as a nutrient source for plants: Litter refers to the organic material, such as leaves, twigs, and dead plant material, that accumulates on the soil surface. It plays a crucial role in providing nutrients to plants. As litter decomposes, it releases essential nutrients like nitrogen, phosphorus, potassium, and various micronutrients into the soil. These nutrients become available for plant uptake, contributing to their growth and development. Litter decomposition also releases organic acids that aid in the breakdown of minerals, making them more accessible to plant roots. Therefore, the presence of litter in the soil serves as a valuable nutrient source for plants. (Warren, R. J., & Wakefield, R. 2016)

B. Role of litter in improving soil structure: Litter also plays a significant role in improving soil structure. The physical properties of soil, such as its texture, aggregation, and porosity, determine its ability to retain water, exchange gases, and provide a suitable environment for root growth. When litter decomposes, it undergoes a process called humification, where organic matter is transformed into stable humus. Humus acts as a binding agent, promoting soil aggregation and the formation of soil aggregates. These aggregates create pore spaces in the soil, improving its structure and allowing for better water infiltration and root penetration. The presence of litter also reduces soil erosion by acting as a protective layer, shielding the soil surface from the impact of raindrops and wind. & Clark, D. A. et al 2007)

C. Litter's contribution to soil organic matter content: Litter significantly contributes to the soil organic matter (SOM) content. SOM consists of various organic materials, including decomposed litter, plant residues, animal waste, and microorganisms. As litter accumulates on the soil surface, it undergoes decomposition through the activity of soil organisms, such as bacteria, fungi, and earthworms. This decomposition process releases carbon compounds into the soil, increasing the SOM content. Higher SOM content enhances soil fertility by improving nutrient retention, moisture holding capacity, and cation exchange capacity. It also supports the growth of beneficial soil organisms and helps maintain soil structure. Therefore, the presence of litter is crucial for sustaining and increasing the organic matter content of the soil, which in turn benefits plant growth and overall soil health (Balvanera, P. et al 2009).

Litter's Water Impact
Water Infiltration and Retention
A. Litter's role in reducing soil erosion.
B. Effect of litter on water infiltration and retention.
C. Litter's impact on mitigating drought and flooding.
A. Litter's role in reducing soil erosion: Litter, which refers to the organic material such as fallen leaves, twigs, and plant debris that accumulates on the soil surface, plays a crucial role in reducing soil erosion. When litter covers the soil, it acts as a protective layer that shields the soil from the impact of raindrops, reducing the force at which raindrops hit the soil surface. This helps prevent soil particles from being dislodged and carried away by runoff. The litter layer also acts as a physical barrier that slows down the flow of water over the soil surface, allowing more time for water to infiltrate into the soil. By reducing the velocity of runoff, litter helps to decrease the erosive power of water and minimizes the likelihood of soil erosion occurring Additionally, the

presence of litter on the soil surface enhances soil structure by promoting the formation of aggregates and improving soil stability. This aids in reducing surface crusting and compaction, which further helps to prevent erosion.

B. Effect of litter on water infiltration and retention: Litter has a significant impact on water infiltration and retention in the soil. The presence of a layer of litter on the soil surface can act as a mulch, creating a barrier that reduces water evaporation and helps to retain soil moisture. The litter layer also regulates soil temperature, reducing the loss of moisture through evapotranspiration.

Moreover, litter enhances water infiltration into the soil by absorbing a portion of the rainwater and gradually releasing it into the soil profile. As rainwater comes into contact with the litter layer, it is intercepted, allowing time for the water to gradually infiltrate into the soil. This process promotes better water distribution and infiltration throughout the soil, reducing the likelihood of surface runoff and increasing the amount of water available for plant roots.

C. Litter's impact on mitigating drought and flooding: The presence of litter on the soil surface can have a positive impact on mitigating both drought and flooding conditions. During droughts, litter acts as a moisture reservoir by reducing water evaporation from the soil surface. The layer of litter helps to conserve soil moisture and prevent rapid drying of the soil, providing a buffer against water loss during dry periods. This can benefit plants by maintaining a more favorable moisture content in the root zone, increasing their resilience to drought stress (Pizer, W 2014).

In terms of flooding, litter plays a role in reducing the impact of heavy rainfall by enhancing water infiltration and reducing surface runoff. When rainfall occurs, the litter layer slows down the flow of water over the soil surface, allowing more time for water to infiltrate into the soil. This helps to reduce the volume and velocity of runoff, minimizing the risk of flash floods and allowing the soil to act as a natural sponge, absorbing and storing the excess water.

Nutrient Cycling and Litter

A. Litter decomposition and nutrient release.
B. Litter's role in maintaining nutrient balance.
C. Influence of litter on soil microbial activity.
A. Litter decomposition refers to the process by which dead organic material, such as fallen leaves, twigs, and dead plants, are broken down and transformed into simpler compounds by decomposers

such as bacteria, fungi, and other microorganisms. During this process, the decomposers consume the organic matter and release nutrients, such as nitrogen, phosphorus, and potassium, back into the soil in a form that can be taken up by plants. This nutrient release is a crucial step in nutrient cycling, as it replenishes the soil with essential elements required for plant growth and ecosystem functioning (Nair, P. K. R. 2004).

B. Litter plays a vital role in maintaining nutrient balance within ecosystems. When plants shed their leaves or other organic material, they contribute to the formation of litter on the forest floor or the ground surface. This litter acts as a reservoir of nutrients, holding and storing them until they are released through decomposition. By accumulating and gradually releasing nutrients, litter helps to regulate nutrient availability in the ecosystem, ensuring that essential elements are not depleted too quickly or become excessively concentrated in the soil. In this way, litter helps to maintain a balanced nutrient cycle and supports the productivity and sustainability of the ecosystem Nair, P. K. R. (2004).

C. Litter has a significant influence on soil microbial activity. Microorganisms, such as bacteria and fungi, are responsible for the decomposition of litter. They break down complex organic compounds into simpler forms, releasing nutrients in the process. The presence of litter provides a food source and habitat for these microorganisms, promoting their growth and activity. Increased microbial activity leads to more rapid litter decomposition and nutrient release. Conversely, the absence of litter can limit microbial activity and nutrient cycling in the soil. Soil microorganisms also interact with plant roots, forming symbiotic relationships that enhance nutrient uptake by plants. Therefore, the presence and quality of litter influence soil microbial communities, which in turn affect nutrient cycling and the overall functioning of ecosystems Nair, P. K. R. 2004).

Biodiversity & Litter Resilience

A. Litter as habitat and food source for soil organisms
B. Role of litter in supporting diverse plant and animal species
C. Link between litter and ecological resilience
A. According to ….litter is a habitat and food source for soil organisms: Litter, which refers to dead organic material such as leaves, twigs, and decaying plant matter, plays a crucial role in promoting biodiversity by serving as a habitat and food source for soil organisms. Soil organisms include a diverse range of microorganisms, fungi, bacteria, insects, earthworms, and other invertebrates. Litter provides a protective cover and shelter for soil organisms, creating a favorable

microenvironment. They went on to stress that many organisms, especially decomposers such as bacteria and fungi, rely on litter as a food source. They break down the complex organic compounds in the litter into simpler forms, releasing nutrients that are essential for the growth of plants. In addition to decomposers, litter also supports a variety of detritivores, which are organisms that directly consume dead plant material. Examples include millipedes, woodlice, and certain species of beetles. These detritivores break down the litter into smaller pieces, facilitating its decomposition and nutrient release. The presence of a diverse community of soil organisms not only enhances nutrient cycling but also improves soil structure, water retention, and overall soil health. This, in turn, promotes the growth of plants and supports a range of above-ground organisms, contributing to overall ecosystem functioning and biodiversity.

B. Role of litter in supporting diverse plant and animal species: Litter plays a crucial role in supporting diverse plant and animal species in ecosystems. As mentioned earlier, litter provides a nutrient-rich environment for plant growth. When litter decomposes, it releases essential nutrients such as nitrogen, phosphorus, and potassium, which are vital for plant development. In forests, for example, fallen leaves create a layer of litter on the forest floor. This litter acts as a natural mulch, providing insulation and moisture retention, which supports the germination and establishment of seedlings. The layer of litter also helps suppress the growth of certain competing plant species and provides a suitable microenvironment for the growth of specific plant communities. Furthermore, the presence of diverse plant species in an ecosystem contributes to overall biodiversity. Different plant species provide different habitat structures, food sources, and niches for various animal species. For example, the diversity of plant species in a forest influences the diversity of bird species, as different birds rely on different plant types for nesting, food, and shelter (Spain, A. V. 2001).

Moreover, some animal species directly depend on litter for their survival. Invertebrates, such as insects and earthworms, use litter as a habitat, nesting material, and a source of food. These invertebrates, in turn, serve as a food source for higher trophic levels, including birds, mammals, and reptiles.

C. Link between litter and ecological resilience: Litter plays a crucial role in enhancing ecological resilience, which is the ability of an ecosystem to withstand and recover from disturbances. The presence of a diverse litter layer contributes to the resilience of ecosystems in several ways.

First, litter acts as a buffer against extreme weather conditions. It provides insulation, reducing temperature fluctuations and preventing soil erosion. During heavy rainfall, litter absorbs and slows down the flow of water, reducing the risk of runoff and flooding. This helps to maintain soil moisture and prevent nutrient loss (Balvanera, P. 2009).

Second, the decomposition of litter releases nutrients slowly over time, providing a steady supply of essential elements for plant growth. This nutrient cycling ensures the long-term sustainability of plant communities and supports the functioning of ecosystems. Furthermore, the diversity of litter composition and structure influences the diversity of soil organisms. A diverse community of decomposers and detritivores increases the efficiency of litter decomposition and nutrient cycling, making the ecosystem more resilient to changes in environmental conditions (Balvanera, P. 2009).

In ecosystems where disturbances occur, such as wildfires or logging, the presence of a diverse litter layer can aid in the recovery process.

Litter Removal Impacts

a. Consequences of excessive litter removal

b. Loss of ecosystem services and productivity

c. Importance of maintaining an adequate litter layer

Consequences of excessive litter removal

Soil erosion: Litter, which consists of leaves, twigs, and other organic materials, plays a crucial role in preventing soil erosion. When excessive litter is removed, the soil becomes more exposed to wind and water erosion. This can lead to the loss of topsoil, decreased soil fertility, and reduced agricultural productivity.

Nutrient cycling disruption: Litter acts as a natural source of nutrients for plants and microorganisms. Excessive removal of litter disrupts the nutrient cycling process, as there is a decrease in the organic matter available for decomposition and nutrient release. This can negatively impact plant growth, soil fertility, and overall ecosystem productivity.

Habitat loss: Litter provides shelter, nesting sites, and protection for a wide range of organisms, including insects, small mammals, and amphibians. Excessive removal of litter can result in habitat loss, which can disrupt the natural balance of ecosystems and lead to declines in biodiversity.

Increased invasive species risk: Litter acts as a protective layer that helps prevent the establishment and spread of invasive plant species. When litter is removed, it creates open spaces and increased light availability, providing opportunities for invasive species to colonize and outcompete native plants.

Loss of ecosystem services and productivity:

Water regulation: The presence of a healthy litter layer helps regulate water flow and infiltration. It helps to reduce surface runoff, allowing water to percolate into the soil and recharge groundwater reserves. Excessive litter removal can disrupt this process, leading to increased runoff, decreased water quality, and the loss of water regulation services.

Nutrient retention: Litter plays a vital role in retaining nutrients within ecosystems. It acts as a sponge, absorbing and holding onto nutrients, preventing their loss through leaching. By removing litter, the capacity of ecosystems to retain nutrients is diminished, leading to nutrient imbalances and reduced productivity.

Carbon sequestration: Litter contributes to the accumulation of organic carbon in soils. Removing excessive litter reduces the input of organic matter, which reduces the capacity of ecosystems to sequester carbon. This can exacerbate climate change by increasing greenhouse gas concentrations in the atmosphere.

Importance of maintaining an adequate litter layer

Nutrient cycling: The decomposition of litter releases nutrients back into the soil, making them available for plant uptake. Maintaining an adequate litter layer ensures a continuous supply of organic matter for nutrient cycling, supporting plant growth and maintaining soil fertility.

Soil moisture regulation: A layer of litter helps regulate soil moisture by reducing evaporation and improving water infiltration. It acts as a mulch, preserving soil moisture and providing a more favorable environment for plant roots (Wakefield, R. 2016).

Erosion control: Litter acts as a natural barrier against soil erosion caused by wind and water. It protects the soil surface, preventing the loss of topsoil and maintaining soil structure.

Habitat and biodiversity support: Litter provides habitat, food, and protection for a variety of organisms, including insects, fungi, and small mammals. By maintaining an adequate litter layer, we support biodiversity and the ecological balance within ecosystems.

Carbon storage: Litter contributes to carbon sequestration by storing organic carbon in soils. By preserving an adequate litter layer, we enhance the capacity of ecosystems to mitigate climate change by capturing and storing carbon dioxide from the atmosphere overall, litter acts as a protective layer, influencing the water cycle by reducing soil erosion, enhancing water infiltration, and improving water retention. These factors contribute to the mitigation of both drought and flooding events, making litter management an important consideration for sustainable land and water management practices (Clark, D. A. 2007).

Conclusion

In conclusion, litter plays a crucial role in agroforestry ecosystems and has significant implications for their overall health and productivity. It serves as a vital source of organic matter, nutrients, and moisture retention, contributing to soil fertility and enhancing the growth of crops and trees. Additionally, litter acts as a protective layer, reducing soil erosion and promoting soil biodiversity.

Given the importance of litter in agroforestry systems, there is a need for further research to better understand its dynamics and the specific mechanisms through which it influences ecosystem functioning. This research could include studying the decomposition rates of different types of litter, investigating the impact of litter quality on nutrient cycling, and exploring the interactions between litter and soil organisms.

Furthermore, promoting sustainable litter management practices is essential for optimizing the benefits of litter in agroforestry. This involves implementing strategies that maintain a balanced and diverse litter composition, such as incorporating different plant species with varying litter qualities. It also includes adopting practices that minimize litter loss through erosion, such as contour plowing or using vegetative barriers.

By investing in further research and promoting sustainable litter management practices, we can enhance the resilience and productivity of agroforestry ecosystems, while also contributing to the broader goals of sustainable agriculture and environmental conservation. It is through these efforts that we can unlock the full potential of litter in agroforestry systems and pave the way for more sustainable and productive agricultural practices.

My Recommendations

1. I call on Governments and other institutions to inculcate agroforestry in the agric policy, environmental management policy and even into school curriculums for agriculture and environment.
2. Farmers should endeavor adopt agroforestry as a way of farming.
3. Conservation practitioners should as much as possible endeavor to teach community members to adopt agroforestry as a farming system.
4. Agroforestry tree species should be used to enrich our home stat, i.e.: people should crop agroforestry tree species e.g. leucaena leucocephala, gliricidia sepium, etc, along their boundaries of their land and edges.
5. People should adopt living fences that are selected for agroforestry species so that it would enrich their soil.

Reference

Ajayi, O. C., Akinnifesi, F. K., Sileshi, G. W., Kanjipite, W., Chakeredza, S., &Mngomba, S. (2009). Soil quality effects of Erythrinaabyssinica and Grevillea robusta planted as live stakes in maize fields on steep slopes in central Malawi. Agriculture, Ecosystems & Environment, 134(3-4), 121-130.

Chazdon, R. L., Harvey, C. A., Komar, O., Griffith, D. M., Ferguson, B. G., Martínez-Ramos, M., ... &Balvanera, P. (2009). Beyond reserves: A research agenda for conserving biodiversity in human-modified tropical landscapes. Biotropica, 41(2), 142-153.

Kumar, B. M., & Nair, P. K. R. (2004). The enigma of tropical homegardens. Agroforestry Systems, 61(1-3), 135-152.

Kuyah, S., Dietz, J., Muthuri, C., Jamnadass, R., &Mwangi, P. (2012). Co-occurrence patterns of trees along gradients of precipitation and soil nutrient status. Journal of Ecology, 100(6), 1461-1468.

Montagnini, F., & Nair, P. K. R. (2004). Carbon sequestration: An underexploited environmental benefit of agroforestry systems. Agroforestry Systems, 61(1-3), 281-295

Nair, P. K. R. (2012). Agroforestry for soil management (2nd ed.). CRC Press.

Tchoundjeu, Z., Ngo-Mpeck, M. L., Simons, A. J., &Minang, P. A. (2006). Productivity and nutrient cycling of Grevillea robusta–based agroforestry systems in the humid lowlands of Cameroon. Agroforestry Systems, 66(3), 213-224.

"Aldy, J & Pizer, W 2014, Comparability of Effort in International Climate Policy Architecture, Discussion Paper 2014-62, Harvard Project on Climate Agreements, Cambridge, Mass."

Blair, J. M., & Parmelee, R. W. (1990). Decay rates, nitrogen fluxes, and decomposer communities of single-species and mixed-species foliar litter. Ecology, 71(6), 1976-1985.

Deacon, L. J., Warren, R. J., & Wakefield, R. (2016). Litter quality mediated soil processes in the restoration of coal fly-ash landfill sites. Ecological Engineering, 86, 237-244.

Lavelle, P., & Spain, A. V. (2001). Soil ecology. Kluwer Academic Publishers.

Lal, R. (2004). Soil carbon sequestration to mitigate climate change. Geoderma, 123(1-2), 1-22.

Wood, T. K., Lawrence, D., & Clark, D. A. (2007). Contrasting effects of different treefall gap disturbances on soil microbial biomass and litter decomposition in a humid tropical forest. Oecologia, 153(3), 597-606.

YOUR KNOWLEDGE HAS VALUE

- We will publish your bachelor's and master's thesis, essays and papers

- Your own eBook and book - sold worldwide in all relevant shops

- Earn money with each sale

Upload your text at www.GRIN.com and publish for free